WHAT WOULD HAPPEN IF...

THE MOON DISAPPEARED?

Written by Izzi Howell

Illustrated by Paula Bossio

WORLD BOOK

www.worldbook.com

I0110890

READING TIPS

This book asks readers to ponder the question *what would happen if the moon disappeared?* Readers will learn about how our moon was formed, how it moves, and how it affects our planet. They will also learn about risks to our moon's future and contemplate what would happen if it disappeared. Use these tips to help readers consider the ripple effects of certain actions and events.

Before Reading

Explain to readers that this book uses cause and effect to show how the moon affects our planet and how human exploration and other events may affect the moon. Cause and effect can help us think about why things happen the way they do. It can also help us think about what might happen in the future. Encourage readers to be on the lookout for examples of a cause and effect structure as they explore what would happen if the moon disappeared.

During Reading

Discuss with readers how some actions and events have multiple causes and others have multiple effects. Explain that it can be tricky to keep all the if/then scenarios straight in our minds, so it can be helpful to create a visual guide. Encourage readers to draw and add notes to their own cause and effect maps like those found on pages 16-17 and 34-35.

After Reading

After finishing the book, discuss with readers how their understandings and opinions of the moon, how it affects our planet, and risks to the moon's future have changed. Additionally, you can have readers respond to the comprehension questions included on page 46 and complete the Chain of Events activity on page 47 to further extend the learning.

Visit **www.worldbook.com/resources** for additional, free educational materials.

There is a glossary of terms on pages 44–45. Terms defined in the glossary are in boldface type that **looks like this** on their first appearance on any spread (two facing pages).

Contents

Our marvelous moon

Our night sky is dominated by the moon. It's the brightest object by far, lighting up the darkness and helping us find our way. The moon is also Earth's closest neighbor in space and our planet's only **satellite.**

However, the moon is so much more than just a night light. It also controls and influences many other aspects of the natural world, including the tides that rise and fall along our coasts, ocean **currents,** and our planet's climate.

I'm kind of a big deal, you know!

Moonlight also affects animal behavior. Both **predators** and **prey** use the light of the moon to look for food. Some animals look for a mate and **reproduce** at the full moon, when the moon's light is brightest.

FUN FACT!

Scientists believe that the moon was created shortly after Earth formed, which was about 4.5 billion years ago!

We often take the moon for granted. It has been by our side pretty much since the creation of our planet, and it will stay there forever, right? Well, yes, probably! But there are some dangers that could put our moon at risk. This loss would have a big impact on our planet. Let's take a look at what would happen if the moon disappeared.

THINK ABOUT IT!

Can you guess what could destroy the moon or make it disappear? Think for a while, and then check your answers on pages 22-25.

Meet the moon!

You probably see the moon up in the sky most nights (and even some days!), but how much do you really know about it?

DID YOU KNOW?

The moon doesn't produce its own light. It just reflects light from the sun!

It's illegal for any nation to own the moon ... or any planet, star, or body in space for that matter!

The distance between Earth and the moon is 238,897 miles (384,467 kilometers).

The moon is the only other body in space that humans have visited, besides Earth.

The myth of the "man in the moon" comes from the pattern of **craters** on the moon's surface, which some people believe resemble a face.

The moon appears to change shape in the night sky because different parts of it are lit up by sunlight at different times. These changes in shape are known as the moon's phases.

A moon is a natural **satellite** that **orbits** a planet or another large object in space, held in orbit by the larger object's **gravity.** Earth isn't the only planet in our solar system with a moon. Mars has two, Jupiter has more than 90, and Saturn has 146! But in this book, we'll just be focusing on Earth's moon—when we say the moon, that's the one we mean!

FUN FACT!

Saturn's moon Titan is larger than the planet Mercury!

I'm a mega moon!

As well as orbiting Earth, our moon also **rotates** on its **axis.** It spins at nearly the same rate as its orbit around Earth. This means that we always see the same side of the moon in the sky. The other side of the moon is sometimes known as the "dark side," but it isn't actually dark there! It was just called that because no one had ever seen it until a spacecraft photographed it in 1959.

I'm not so mysterious after all!

Scientists believe that the moon was formed when a massive, planet-sized object collided with Earth—crash! This sent a cloud of vaporized rock flying off Earth's surface and up into **orbit** around Earth. The cloud cooled and condensed into a ring of small, solid bodies, which then came together to form the moon.

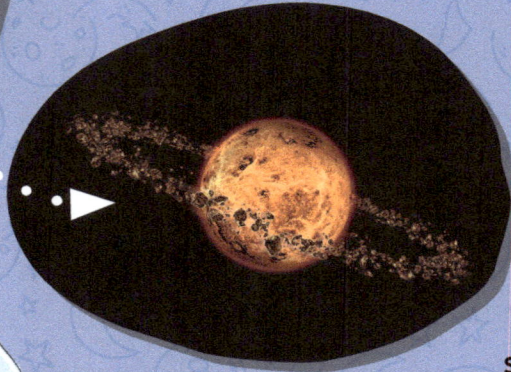

The moon has been a source of interest and fascination since ancient times. Prehistoric people tracked the phases of the moon and used them to create the first calendars. Ancient scientists observed the moon and tried to understand the science behind its movements. When such new instruments as telescopes were invented hundreds of years ago, people were able to create the first maps of the moon's surface.

FUN FACT!

Many traditional religions and mythologies had moon gods and goddesses, including the ancient Roman goddess Luna, the Chinese goddess Chang'e, and the ancient Egyptian god Thoth.

It wasn't until 1969 that humans set foot on the moon as part of the Apollo 11 landing. More astronauts visited the moon over the next few years up until their last trip in 1972. Since then, there have been many missions that have sent uncrewed landers and rovers to explore the moon's surface and collect data. However, there are plans to send people to the moon again, so astronauts may be heading back there before long!

The U.S. astronaut Buzz Aldrin walks on the moon as part of the Apollo 11 landing in 1969.

Imagine that you could travel to the moon and take a walk on its surface. The experience would be totally out of this world! Beneath your feet, the ground would be rocky and dusty. Up above, the sky would always be black, even during the day, giving you a 24-7 view of the stars. You'd also need to carry breathing equipment, because there wouldn't be any air for you to breathe!

FUN FACT!

Moon dust smells like gunpowder, according to astronauts who have visited the moon!

You'd also feel quite different during your moonwalk! The moon has much less mass than Earth, which means that there is less **gravity.** This makes objects fall to the ground more slowly than on Earth, which would make you feel as if you weigh less!

As you made your way across the moon, you might come across massive **craters.** The moon has barely any atmosphere to stop **asteroids** or meteoroids, so many make it down to the moon's surface, where they make dramatic crash landings.

???
DID YOU KNOW?
Some parts of the moon are also covered in dark basalt rock from ancient volcanoes.

This crater is 14 miles (22 kilometers) wide!

Scientists have recently confirmed that water does exist on the moon! However, unfortunately, you wouldn't be able to use it to refill your water bottle. The water on the moon isn't liquid—it's mostly trapped in ice or minerals in the soil.

THINK ABOUT IT!

What do you think you'd enjoy about a walk across the moon? What wouldn't be so fun?

Tides

Even though the moon has far less gravitational pull than Earth, its **gravity** still reaches our planet. The moon's pull on our oceans creates tides—a daily rise and fall in water levels.

Gravity from the moon pulls most on the area nearest the moon and the area on the opposite side of our planet. This creates a bulge in the ocean, which causes a high tide. The places in between these two areas experience low tide. These areas move as Earth **rotates** on its **axis,** which makes tides rise and fall throughout the day in different parts of the world.

Teamwork makes the dream work!

FUN FACT!

Tides are mostly caused by the moon's gravity, but the sun's gravity helps out as well!

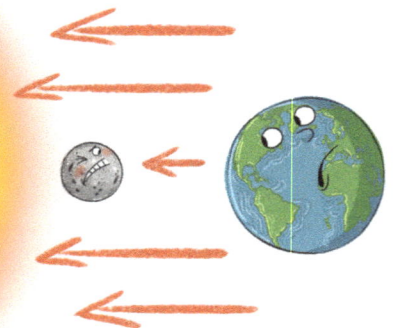

Tides are easiest to spot along the coasts. At high tide, the water level rises and covers more of the shore. At low tide, the water level falls to its lowest point, and more land is left uncovered.

Time to catch some fish!

SS681

At low tide, the boats in this harbor are trapped on the sand, but they can sail away freely at high tide!

DID YOU KNOW?
The moon's gravity also holds Earth in a tilted position as it **orbits** the sun—you'll find out later why this is so important!

THINK ABOUT IT!

Would you rather visit a beach at high tide or low tide? Why?

As the water level drops along the coast, a whole new world is revealed! Welcome to the magical **intertidal zone**—the area of the shore that is uncovered at low tide but underwater at high tide.

Life is never boring in the intertidal zone. At high tide, it's covered by salty seawater and connected to the ocean. As the tide goes out, it's battered by waves. At low tide, it's totally exposed to the air, and the only saltwater that remains is found in tide pools. Luckily, the animals and plants that live in this **habitat** are perfectly adapted to these conditions.

Barnacles use their own natural glue to stick themselves to rocks or even other animals with hard shells, such as crabs or mussels!

Get off me!

Living in the intertidal zone also has its benefits! At high tide, tide pools are refilled with fresh ocean water and living things, providing constant food, **nutrients,** and oxygen for their residents. At low tide, animals in the tide pools are safe from larger ocean **predators,** such as fish, that are left behind in the sea.

Tee-hee, no one will find me here!

Lunchtime! This ruddy turnstone has found a crab to eat in the intertidal zone. Many birds come here at low tide to feast on shellfish, worms, and seaweed.

15

TIDES

What have tides ever done for you (other than destroy your sandcastle!)? Well, quite a lot, actually! Tides perform many important jobs, both in the natural world and for people. Let's take a look.

TIDES ARE TERRIFIC!

CURRENTS

Tides contribute to ocean **currents,** which move water around the world's oceans. They carry warm water from tropical areas toward the poles, which helps to warm up colder areas and cool down warmer ones! This helps stabilize Earth's climate.

RESHAPING COASTLINES

Over time, the constant movement of water erodes (wears away) rock and sand along the shore. Tiny pieces of sand and rock are carried by the waves and dropped in other areas, creating new beaches and sandbars (ridges of sand in the water).

ENERGY

Tides can be used to produce energy! The rise and fall of the sea level pushes **turbines** in tidal power plants. These turbines are connected to generators that produce electricity.

DID YOU KNOW?

Tidal power is a **renewable** source of energy. It will never run out, as long as we still have tides!

SHIPS

Ships can't sail into some shallow areas at low tide because the water isn't deep enough for them to stay afloat. High tide to the rescue! When the water level rises, ships can sail into these areas that would otherwise be inaccessible.

Come on, tide!

Life by moonlight

The moon also has a big impact on animals and their behavior, especially **nocturnal** animals. Nocturnal animals, such as owls, bats, and foxes, are active during the night and sleep during the day.

Many **predators** are nocturnal. They hunt at night, hiding away in the darkness until the time is right to strike! These animals have excellent night vision and need only a small amount of moonlight to spot their **prey.**

THINK ABOUT IT!

Would you like to be nocturnal? How would it affect your life if you were awake during the night and asleep during the day?

FUN FACT!

Owl eyes are so large that they can't move in their sockets! However, their super size lets in more light, which improves the owl's vision!

However, it's not just predators out at night! Nocturnal prey animals also use the darkness to their advantage and hide in the shadows while they are gathering food. Many of these animals don't have great eyesight—they rely instead on their senses of smell, hearing, taste, and even touch to find their way at night!

Smells like trouble!

Hi! I'm Jeff Shima, an American ecologist with a particular interest in fish. During a recent research project studying common triplefin fish, I discovered that their babies grow much faster on bright nights with plenty of moonlight than on dark, cloudy nights. We think this might be because the moonlight helps them find plankton to eat. The predators of common triplefin are also more likely to hide away on bright nights, for fear of running into their own predators!

Eat up, fishies!

LIFE BY MOONLIGHT

For some animals, the moon is a little bit like a natural clock! They time special events, such as mating or **migration,** to coincide with changes in moon phases.

If you were to visit the Great Barrier Reef, off the coast of Australia, in November, just after a full moon, you might be lucky enough to witness an amazing sight. All the coral in the reef (which is the largest system of coral reefs in the world) release eggs to be fertilized at the same time! They use the full moon to **synchronize** with each other. **Spawning** at the same time makes it more likely that the eggs will become fertilized, which means lots more new coral for the reef!

Those tiny pink flecks are coral spawn!

Barau's petrel birds also look for a full moon to know when it's time to head off on their migratory trips to their mating sites. They live near the equator, where there's little difference in season or day length throughout the year, so changes in the moon are a much better way of tracking the passing of time. Just as with coral, birds that arrive at roughly the same time find it much easier to mate.

It's time!

Even after mating, the moon is still useful for some animals. Newly hatched sea turtles use the reflection of the moon in the ocean to find their way from the beach to the water.

No more moon?

OK, so the moon is important ... but is it really at risk? The good news is that it's highly unlikely the moon will ever disappear. It's been by our side for about 4.5 billion years (pretty much since the birth of our planet!), and it will almost certainly stay there for millions more years to come. However, there are a few extreme incidents that could put our moon in danger.

One possible threat could be a massive **asteroid** collision. If a huge rock from outer space collided with the moon, it could damage it or destroy it entirely. Not only would we lose our moon, but Earth would also probably be hit by large pieces of moon rock following the explosion, which would have disastrous consequences for our planet. So, losing the moon would probably be the least of our worries!

There are at least 1.3 million asteroids out there! Most of them are found in a massive belt between Saturn and Jupiter, but they can travel, so watch out!

With plans for human **colonization** of the moon and talk of future mining projects to extract **resources** from it, there's also a risk that humans might end up destroying the moon! The moon contains many valuable natural resources that companies could make a lot of money by mining and selling. Considering that Earth has been seriously harmed by the overextraction of its own resources, it's an understandable worry that we'd treat the moon in the same way. We might end up severely damaging the moon through overextraction or even mining it all away!

THINK ABOUT IT!

Who do you think should be responsible for deciding how humans treat the moon? It could be scientists, politicians, business owners, or someone else. How might their opinions vary?

23

NO MORE MOON?

You might have heard that the moon used to be closer to Earth. It is now moving away at the speed of about 1.5 inches (3.8 centimeters) a year. Should we be worried? Could it drift away altogether?!

Keep in touch!

FUN FACT!

The moon moves away from Earth at around the same speed that your fingernails grow!

In short, no! The moon will never drift away altogether. Scientists estimate that it will reach its maximum distance from Earth in about 50 billion years. At that point, it will take about 47 days to **orbit** Earth, compared to roughly 27 days today.

After that, the moon will start moving back toward Earth. After another 50 billion years or so, it will be so close to our planet that it will be pulled apart by our **gravity.** This sounds scary, but don't worry—we won't be around to see any of this playing out!

The sun is only supposed to have enough fuel for another 5 billion years, at which point it will start to expand and might swallow up Earth. Even if the sun doesn't reach our planet, its heat would destroy our planet. So, there's not much point worrying about the comings and goings of the moon!

THINK ABOUT IT!

Can you think of anything else that might threaten the moon? It can be a serious idea or a silly idea!

It's highly unlikely that the moon will ever disappear, so it's not worth worrying about. However, it is very interesting to consider what our world would look like without it. Read on to find out more!

Tidal trouble

The loss of the moon would have a huge impact on Earth's tides. Tides are mostly created by the moon's **gravity,** so if the moon disappeared, tides would get much smaller and weaker. However, gravity from the sun would still pull on Earth, leaving us with tides about one-third of their current size.

Don't worry, you've still got me!

As we've seen before, tides perform many important roles. If the moon disappeared, we'd not only be left without tidal power or the ability to sail into shallow tidal areas, but we'd also notice a huge impact on our climate.

Without strong tides contributing to ocean **currents,** areas of warm and cold water would stay where they are and wouldn't have the chance to mix. We'd lose the stabilizing effect that ocean currents bring to our planet. As a result, temperatures around the world would probably become much more extreme.

It's a bit chilly, even for me!

Hi! I'm Maria Luneva. I'm an ocean scientist with a particular focus on the circulation of seawater. In one research project, my team and I looked at how tides affect sea ice in the Arctic. We found out that tides are actually responsible for a small decrease in sea ice, because they carry in warm water, which melts the ice.

Ocean currents also bring different kinds of weather with them. Cold currents result in cooler, drier weather, while warm currents lead to hotter, wetter weather. Without currents carrying these patterns of weather around the world, we might notice big changes to our normal weather systems.

Where's the rain I was promised?!

27

TIDAL TROUBLE

If tides became much weaker, the **intertidal zone** would be in big trouble. Water wouldn't reach such a high point at high tide and so wouldn't refill many tide pools. Large numbers of tide pools would dry out, making much of the **habitat** totally uninhabitable for many intertidal species.

Many intertidal species would struggle to survive with no tides bringing in fresh food and water, and fewer tide pools in which to shelter. There would be big competition for the remaining tide pools. Many animals would go hungry, and the **populations** of many species would fall.

But wait, couldn't intertidal animals just move away from the intertidal zone and go deeper into the sea? After all, they live there part-time anyway! It would be a simple solution for some intertidal animals to switch to full-time life in the sea, since they'd have water and food 24-7.

While a move to the sea would save some intertidal species, it would also create new problems. The arrival of more intertidal species might unbalance the ocean **ecosystem.** For a start, there would be more hungry mouths to feed, resulting in less food for some species. Too many **predators** or too many **prey** species would also be hard to sustain. But in the end, the new ocean ecosystem would balance itself out.

FUN FACT!

Sea anemones look like plants, but they are actually predatory animals often found in tide pools! They eat small fish, crabs, and plankton ... anything their tentacles can grab, really!

THINK ABOUT IT!

What effect do you think the shrinking of the intertidal zone would have on seabirds and other animals that come there to look for food?

I can't find any lunch!

A darker night

The loss of the moon would make a huge and obvious change to the night sky. Without the moon reflecting light from the sun, nighttime would be much, much darker!

Animals that depend on moonlight might find it hard to adjust to the darkness. Some **predators** that hunt at night would struggle to spot their **prey. Nocturnal** prey species might also find it harder to spot predators creeping up on them in the dark. Some might struggle to survive, and their **population** would fall.

Where did you come from?!

Nocturnal animals that use multiple senses to find food and keep safe would probably find it easiest to adjust. They'd be able to hear, smell, or even taste their surroundings instead, even if they couldn't see them.

Who needs light when you can see heat?!

FUN FACT!
Lions prefer to hunt during darker nights because their prey is less likely to see them coming. So lion numbers might increase without the moon around!

Hi! We're Catarina Rydin and Kristina Bolinder. We're biologists with a particular interest in plant pollination. One of our most exciting discoveries was a plant that depends on the moon to help with its pollination! At a full moon, the plant releases tiny see-through droplets that glitter like diamonds in the moonlight. They attract insects to the plant for pollination. Without the moon, this plant would have no chance of being pollinated, because the insects just wouldn't be interested!

Flashlights would become essential in dark rural areas!

The loss of moonlight would also create safety issues in areas without artificial lights. People would trip and fall more often because they wouldn't be able to see where they are stepping. Luckily, this would be a fairly simple problem to solve. We could build more lampposts, or people could get into the habit of carrying flashlights with them at night.

FUN FACT!

Some dung beetles use moonlight as a compass to help them find the safest place to bury dung balls!

Species that use the moon like a clock would experience major confusion. Without moonlight to help them **synchronize**, they might be ready to **reproduce** at different times. This would make mating much harder, because fewer animals would be available to mate at the same time. This would result in fewer babies born, and so their **populations** would decrease.

Many animals that live in groups also choose to mate at the same time because there's safety in numbers. It's easier for them to protect their vulnerable babies from **predators** when there are lots of adults around to spot danger and fight back. If these species spread out their mating throughout the year, their young would be at greater risk of attack.

Grunion fish time their mating around the tides (which are controlled by the moon). When their young hatch on the beach, they are washed back out to sea by the high tide.

THINK ABOUT IT!

How much do you think you would be affected by the loss of the moon, compared to a **nocturnal** animal?

It's definitely worse for me!

If life without the moon became too challenging, some **nocturnal** species might slowly become **diurnal** (active during the day) and switch over to daytime living. Let's look at the consequences of this change.

What would happen if nocturnal animals became diurnal?

In deserts, many animals are nocturnal to avoid the heat of the sun. They wouldn't be able to come out during the day, since it would be far too hot for them!

Some animals would stay put in hot **habitats** and adapt to a nocturnal life without the moon.

Some animals might move to a cooler area more suitable for daytime living.

Saw-scaled vipers beat the heat by coming out at night to hunt for scorpions, lizards, and birds.

At the moment, many ecosystems have different sets of **predators** and prey that are active during the day and at night. These species don't compete for food, because they are never awake at the same time! However, if nocturnal species became diurnal (active during the day), they would suddenly come face-to-face with daytime species.

If diurnal and nocturnal species competed for food, more effective predators might end up dominating and taking all the food.

Weaker species would miss out on food, and they might starve. Their population would decrease.

Major population changes would upset the fragile balance of the ecosystem. However, over time, things would readjust, and a new normal would be found.

DID YOU KNOW?

Hawks and owls can live in the same habitat without competing for food because hawks are diurnal and owls are nocturnal!

THINK ABOUT IT!

What do you think would happen to hawk and owl populations if owls became diurnal? How might prey populations be affected?

I take the day shift ...

... and I take the night shift!

Bye-bye seasons?

Do you remember reading way back on page 13 that the moon's **gravity** holds Earth in a tilted position as it **orbits** the sun? Let's dig in and find out why this is important, and what would happen if the moon wasn't there to keep Earth tilted!

The moon's gravity holds Earth in a steady tilt about 23.5 degrees to the side, relative to its orbit of the sun. Because of this tilt, Earth's Northern and Southern **Hemispheres** tilt toward or away from the sun at different times of the year. This is why we experience seasons.

No one really knows exactly how Earth's tilt would change without the moon to hold it in place. We might end up with no tilt, which would result in no seasons at all, or a much greater tilt, which would make the high and low temperatures of each season more extreme. Earth's tilt could even change from time to time, making our seasons very unpredictable and confusing!

The Northern Hemisphere is starting to tilt toward the sun and experiences spring. The Southern Hemisphere is starting to tilt away from the sun and experiences fall.

The Northern Hemisphere is tilted away from the sun and experiences winter. The Southern Hemisphere is tilted toward the sun and experiences summer.

The Northern Hemisphere is tilted toward the sun and experiences summer. The Southern Hemisphere is tilted away from the sun and experiences winter.

The Northern Hemisphere is starting to tilt away from the sun and experiences fall. The Southern Hemisphere is starting to tilt toward the sun and experiences spring.

DID YOU KNOW?
Tropical areas near the equator don't experience four different seasons. They usually have one or two dry and wet seasons each year.

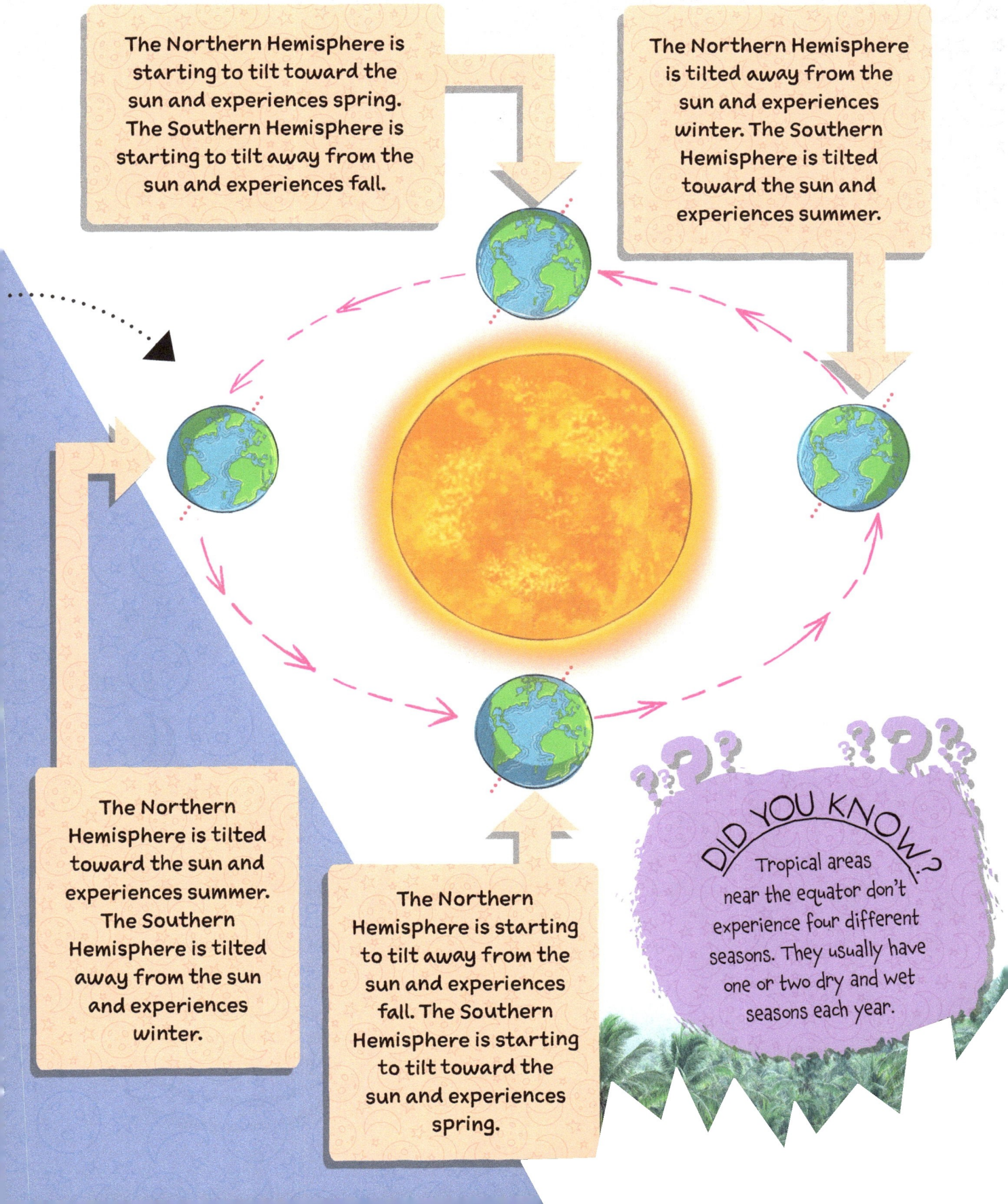

BYE-BYE SEASONS?

Earth's tilt does gradually change by itself. In the past, it has had a slightly larger and a slightly smaller tilt than it has today. These changes affected Earth's climate and contributed to extreme climate events, such as **ice ages.** This is when most of the planet is covered in ice. Scientists believe that the loss of the moon could plunge Earth into another ice age. Brrr!

DID YOU KNOW?
Earth's tilt was at its maximum size about 10,000 years ago. In another 10,000 years, its tilt will be at its smallest size.

Earth's most recent ice age ended about 11,500 years ago. Woolly mammoths like us were still alive back then! Our long hair kept us warm.

Luckily, Earth's tilt wouldn't change immediately following the loss of the moon. Earth has a lot of momentum, and so it would take a while for its tilt to shift. However, we would gradually experience significant changes to our seasons. Let's see how that would affect life on Earth.

The life cycles of many animals follow the seasons. Without our normal seasons to guide them, baby animals might be born at the wrong time of year, when there is less food available.

Many **crops** that we depend on for food would struggle to grow in the different seasons. This could create problems with our food supply, and people might go hungry.

Many wild plants might also find it hard to grow in the new conditions. As a result, wild animals might starve, and their **populations** would suffer.

THINK ABOUT IT!

Would you rather have no seasons or extreme seasons? Why?

Conclusion

Next time you go outside at night, say a quick thank you to our marvelous moon! Not only does it brighten our nights, but it also controls our tides, seasons, and the behavior of many different animals!

In the highly improbable event that we did lose the moon, it would have a massive impact on life on Earth. Our tides would be greatly reduced, many animals would struggle to adjust to the darkness, and we'd experience major changes to our seasons.

DID YOU KNOW?
If the moon disappeared, we'd never experience a solar or lunar eclipse ever again.

These extreme seasons aren't so bad after all!

However, the loss of our moon wouldn't be the end of life on Earth. Many species would struggle to adjust their behavior and habits, and some might die out. But in the end, most animals that lived in sync with the moon would find a new way to survive. Humans would also adapt to the new climate, seasons, and weather. Our night sky would be much darker, but we'd be alright in the end!

FUN FACT!

Without the moon, we'd be able to see more stars from Earth!

Summary

So, exactly what would happen if the moon disappeared? Check your understanding of the information in this book.

The moon disappears because of an **asteroid** collision or terrible accident.

Earth is no longer held in a tilted position as it **orbits** the sun. Eventually, it will settle into a new position.

Without **gravity** from the moon, tides are only one-third of their original size and strength.

The night sky becomes much darker.

Ocean **currents** become less powerful, which results in less mixing of warm and cold water in the oceans. This could make our climate more extreme.

Conditions in the **intertidal zone** change considerably as many tidal pools are no longer refilled at high tide. Many animals there would struggle to survive without food and shelter.

Some intertidal animals may move into the ocean. This may create issues for the animals that already live there.

If Earth loses its tilt, we will lose our seasons.

If Earth's tilt becomes larger, our seasons may become more extreme.

Some nocturnal animals may struggle to find enough food without moonlight to guide them.

Animals that use the moon as a clock to **synchronize migration** and reproduction would find it much harder to time these events correctly. They may struggle to mate, and their populations may fall.

Eventually, animals will adjust to their new life, and **ecosystems** will find a new balance.

THINK ABOUT IT!

What do you think you'd miss most about the moon if it disappeared? Why?

To survive, some nocturnal animals could become **diurnal.**

Rise and shine!

The change in routine in previously nocturnal species may upset existing ecosystems, and populations may change as a result.

Glossary

asteroid—a very large rock that orbits the sun

axis—a real or imaginary line through the center of an object that is spinning

colonization—settling somewhere else (in this case, space!)

crater—a large hole in the ground

crop—a plant grown for food, such as apples, carrots, or potatoes

current—the movement of water in a particular direction

diurnal—describes an animal that is active during the daytime and rests at night (like you, probably!)

ecosystem—all of the living things in an area and the relationship between them

gravity—the force that attracts objects toward each other

habitat—the place where an animal or plant usually lives

hemisphere—one of two halves of Earth (usually split around the equator)

ice age—a time when temperatures on Earth are much colder than usual, and large ice sheets and glaciers cover lots of the surface

intertidal zone—the area along the coast that is covered by water at high tide and left uncovered at low tide

44

migration—when an animal travels to a new place, usually as the season changes

nocturnal—describes an animal that is active during the night and rests during the day

nutrient—something that living things need in order to grow

orbit—to follow a curved path around a planet or a star

population—how many animals or plants of the same type live in an area

predator—an animal that kills and eats other animals for food—watch out!

prey—an animal that is killed and eaten by other animals

renewable—something that can be used again and again and will never run out

reproduce—to produce new, young animals or plants

resource—a useful material

rotate—to turn in a circle

satellite—something that moves around a larger object in space

spawn—to lay eggs

synchronize—to make happen at the same time

turbine—a machine for making electricity with a wheel that's turned by flowing air, steam, or water

Review and reflect

COMPREHENSION QUESTIONS

Meet the moon!
- Why do we always see the same side of the moon in the sky?
- How was the moon formed?

Tides
- How does the moon create tides in Earth's oceans?
- What are some of the important "jobs" that tides perform?

Life by moonlight
- How is the moon helpful to some animals?
- What are some nocturnal animals? How are these animals equipped for life at night?

No more moon?
- What are some of the extreme incidents that could put our moon in danger? Why?
- Will the moon ever drift away from Earth altogether? Why or why not?

Tidal trouble
- How would the loss of the moon impact Earth's tides?
- How would this impact on tides affect some animals?

A darker night
- How would a darker night affect some animals?
- What could people do to adjust to a darker night?

Bye-bye seasons?
- How does the moon affect seasons?
- How would changes to our seasons impact life on Earth?

Conclusion and summary
- After reading this book and considering what would happen if the moon disappeared, what is your biggest takeaway? Why?

MAKE A CHAIN OF EVENTS!

Creating a paper chain can help you explore and visualize how cause and effect relationships can be thought of as a sequence of events.

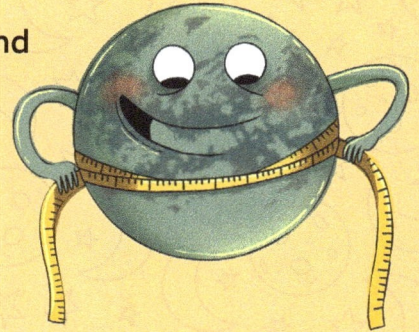

You'll need:
- Pencil
- Scratch paper
- Pens or markers
- Stapler and staples
- Strips of paper (2 colors, if possible)

Instructions:

1. **Select a focus:** Choose a specific aspect from the book that caught your attention—it could be how the moon was formed or how some animals would be affected by the loss of the moon.

2. **Brainstorm causes and effects:** On a sheet of scratch paper, brainstorm and list the causes and effects related to your chosen focus. Think critically about the factors that contributed to or resulted from your focus. You can always look back in the text for ideas!

3. **Write on strips:** Write each cause and each effect on its own strip of paper. If you have different colored paper, use one color for the cause strips and the other for the effect strips.

4. **Create the paper chain:** Organize your strips into causes and effects. Start forming a paper chain to show how a cause leads to an effect. Use the stapler to connect the two strips. Continue adding cause and effect strips as links in your chain. When you've finished, you should be able to start at the beginning of your chain and read through each chain link in a logical order.

5. **Linking multiple chains:** If your focus has multiple causes or effects, you can create additional chains and link them together to show how complex cause and effect relationships can be!

Write about it!

Look at the paper chain you created and how the causes link to effects (which in turn link to other causes!). How might breaking a link in the chain impact the overall sequence of events?

World Book, Inc.
180 North LaSalle Street
Suite 900
Chicago, Illinois 60601
USA

For information about other World Book publications, visit our website at www.worldbook.com or call 1-800-WORLDBK (967-5325).

For information about sales to schools and libraries, call 1-800-975-3250 (United States), or 1-800-837-5365 (Canada).

© 2025 by World Book, Inc. All rights reserved. No part of this publication may be reproduced, stored in a retrieval system, or transmitted in any form or by any means (electronic, mechanical, photocopying, recording, or otherwise) without written permission from World Book, Inc.

WORLD BOOK and the GLOBE DEVICE are registered trademarks or trademarks of World Book, Inc.

Library of Congress Control Number: 2024941782

What Would Happen If...
ISBN: 978-0-7166-7125-1 (set, hard cover)

The Moon Disappeared?
ISBN: 978-0-7166-7129-9 (hard cover)
ISBN: 978-0-7166-7141-1 (e-book)
ISBN: 978-0-7166-7135-0 (soft cover)

Staff

Editorial

Vice President
Tom Evans

Editorial Project Coordinator
Kaile Kilner

Curriculum Designer
Caroline Davidson

Senior Editor
Shawn Brennan

Proofreader
Nathalie Strassheim

Graphics and Design

Senior Visual
Communications Designer
Melanie Bender

Digital Asset Specialist
Rosalia Bledsoe

Written by Izzi Howell
Illustrated by Paula Bossio

Developed with World
Book by White-Thomson
Publishing LTD

Acknowledgments

4-5 © Klagyivik Viktor/Shutterstock; © Miguel Regalado, Shutterstock
6-7 © Korionov/Shutterstock; © Daniel Fung, Shutterstock
8-9 © Jacques Dayan, Shutterstock; NASA
10-11 NASA; NASA/Goddard Space Flight Center/Arizona State University
12-13 © Jason Batterham, Shutterstock; © Gordon Bell, Shutterstock; © Peter Hermes Furian, Shutterstock
14-15 © Brian Lasenby, Shutterstock; © Emily Duwan, Shutterstock
16-17 © Stock for you/Shutterstock; © William C. White, Shutterstock
18-19 © Panther Media GmbH/Alamy Images; © Danita Delimont, Shutterstock
20-21 © Ricardo Reitmeyer, Shutterstock; © Naoto Jack Fukushima, Shutterstock
22-23 © Alexyz3d/Shutterstock; © Dima Zel/Shutterstock
24-25 © muratart/Shutterstock; © Elena11/Shutterstock
26-27 © Novikov Alex, Shutterstock; © Aphelleon/Shutterstock
28-29 © SandyHappy/Shutterstock; © Spiroview Inc/Shutterstock
30-31 © Agami Photo Agency/Shutterstock; © Sean Nel, Shutterstock
32-33 © Joachim Bago, Shutterstock; © Michele and Tom Grimm, Alamy Images
34-35 © Sheril Kannoth, Shutterstock
36-37 © Tonio_75/Shutterstock
38-39 © Bachkova Natalia, Shutterstock; © ABCDstock/Shutterstock; © Bobkov Evgeniy, Shutterstock; © Marek Mierzejewski, Shutterstock
40-41 © vovan/Shutterstock; © MISHELLA/Shutterstock

www.ingramcontent.com/pod-product-compliance
Lightning Source LLC
Chambersburg PA
CBHW060857090426
42737CB00023B/3482